HEALING PLANTS, PLANTS TO FEAR, REVISITING BOTANICAL CLASSIFICATIONS

Camilia MacPherson, Ph.D., D.Th.
2016

INTRODUCTION
Every page has to be looked at from every angle and varying distances.

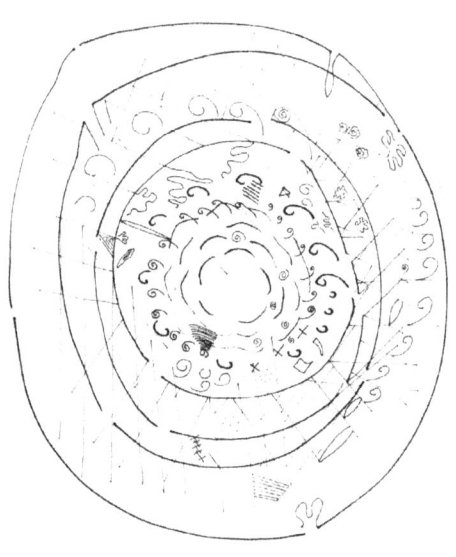

ISBN-13:9781530948239
ISBN-10:1530948231
Email: tamaracpublishers@icloud.com

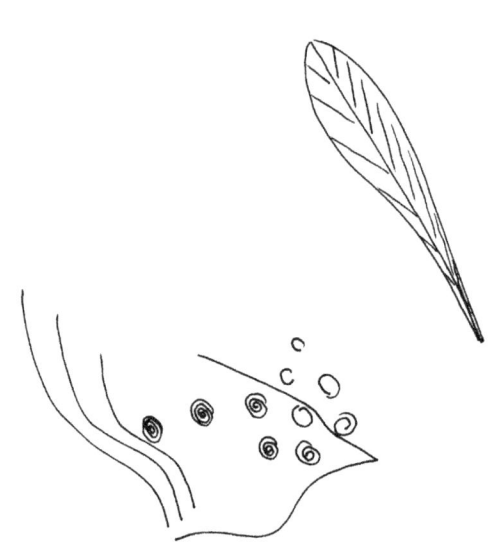
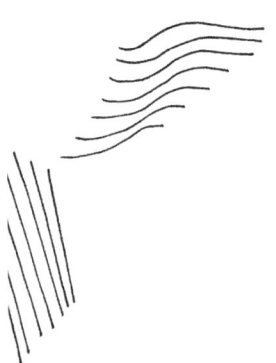

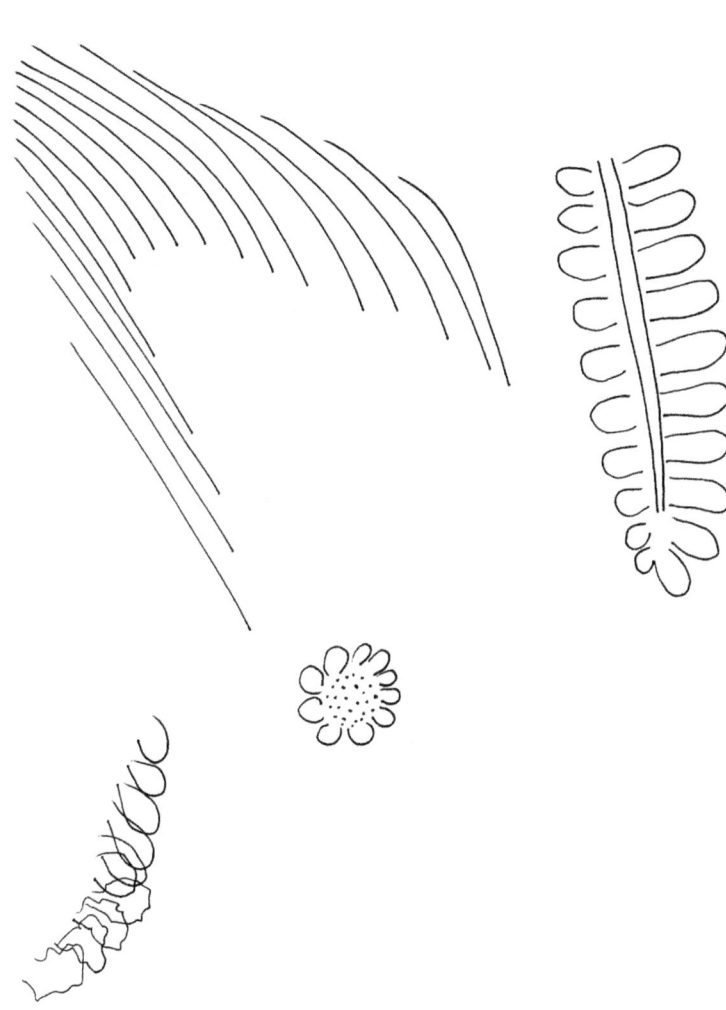

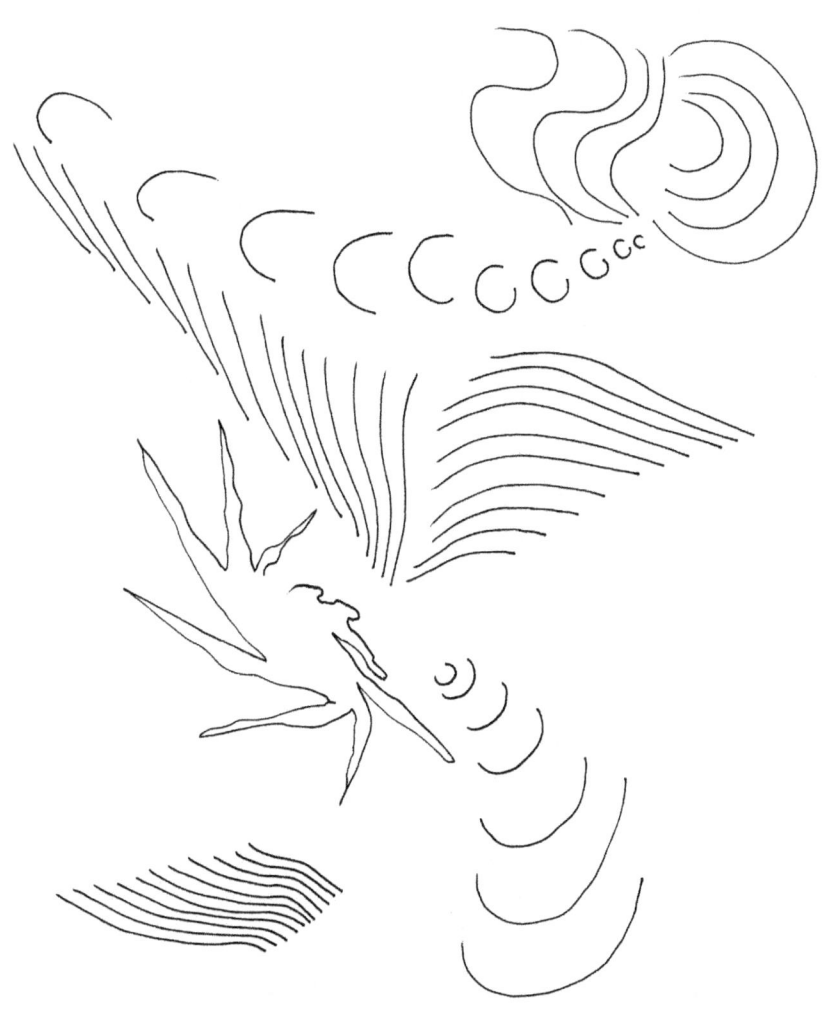

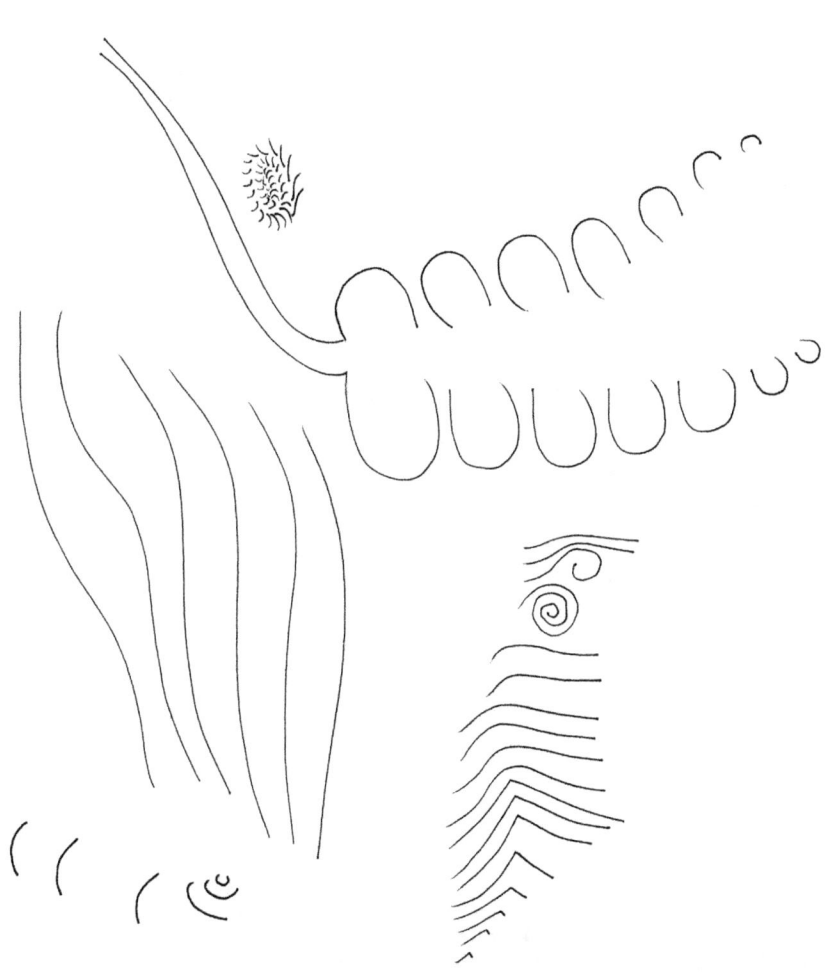

PLANTS TO FEAR

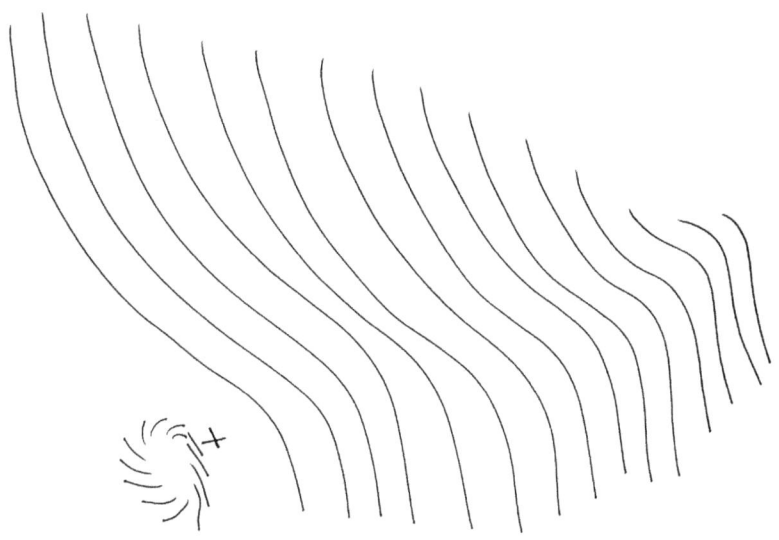

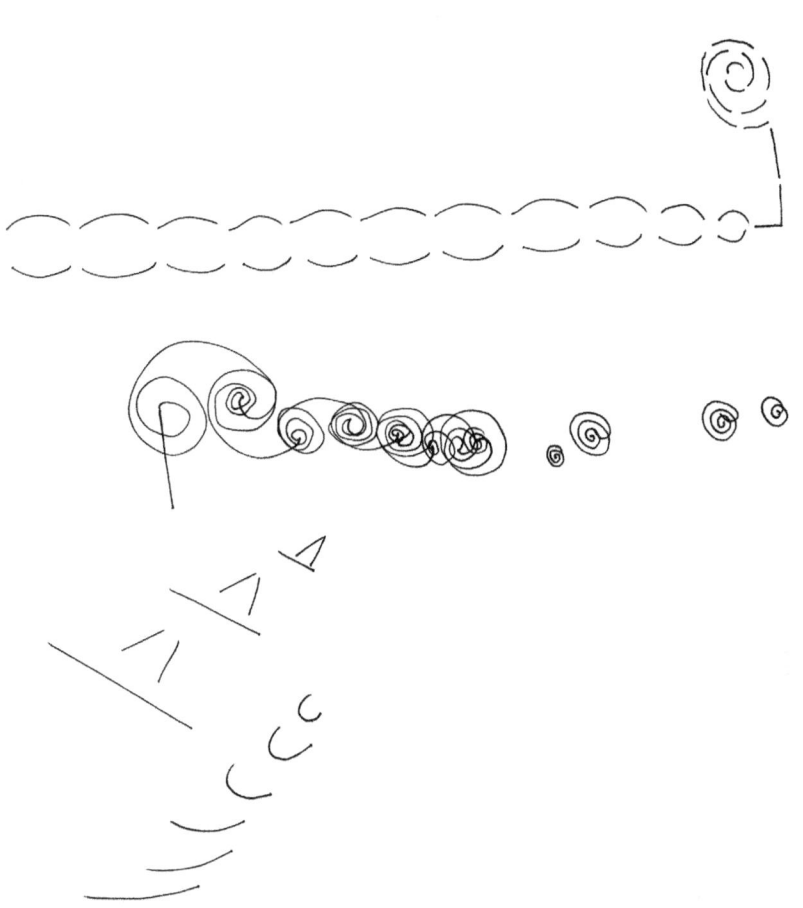

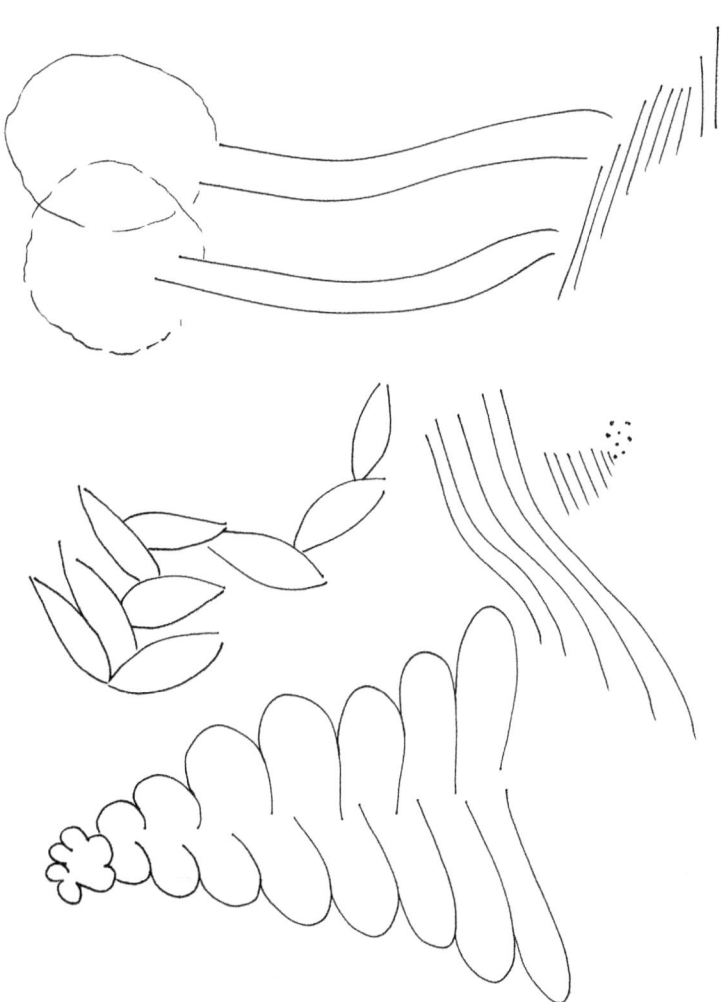

REVISITING BOTANICAL CLASSIFICATIONS

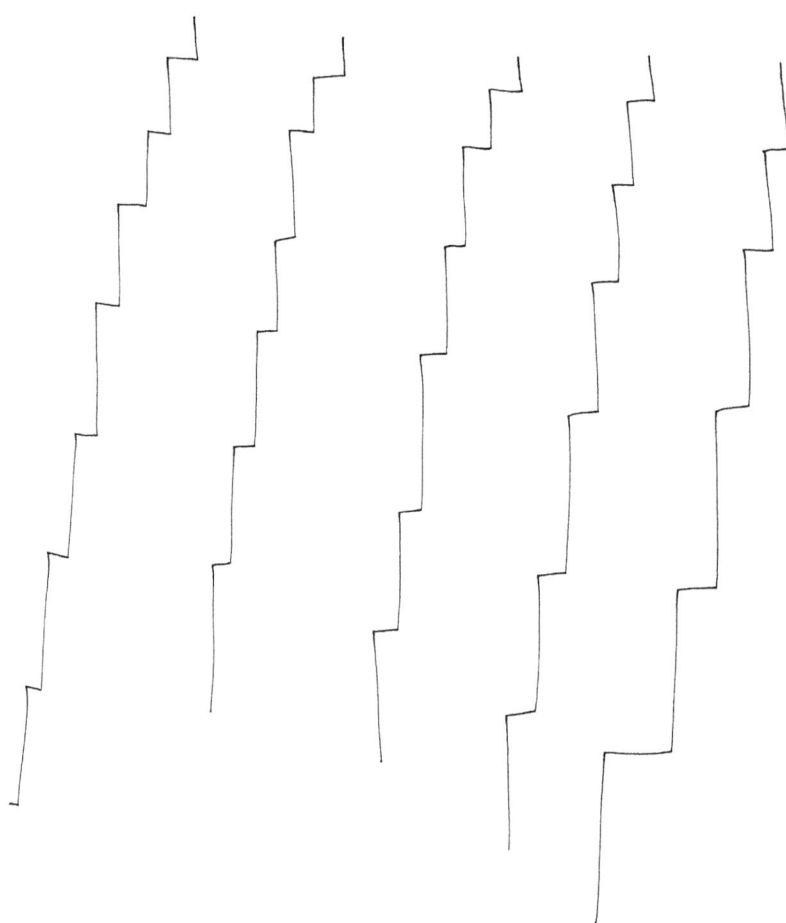

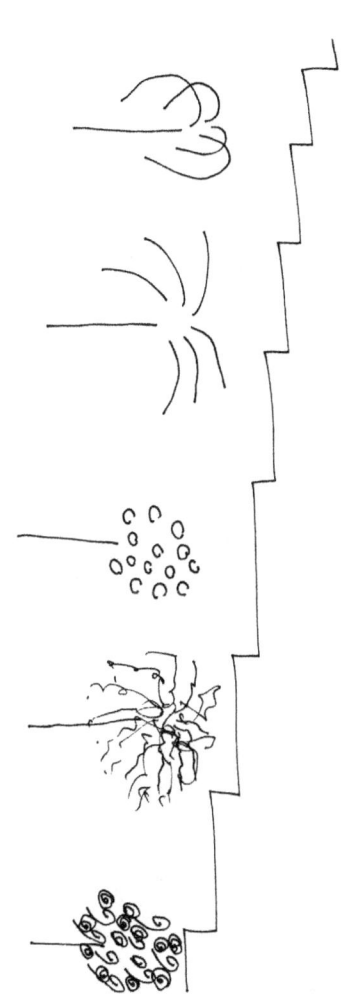

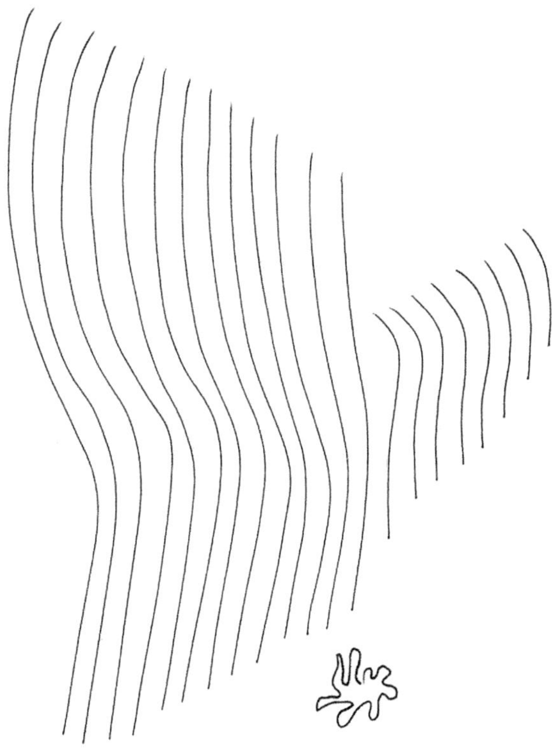

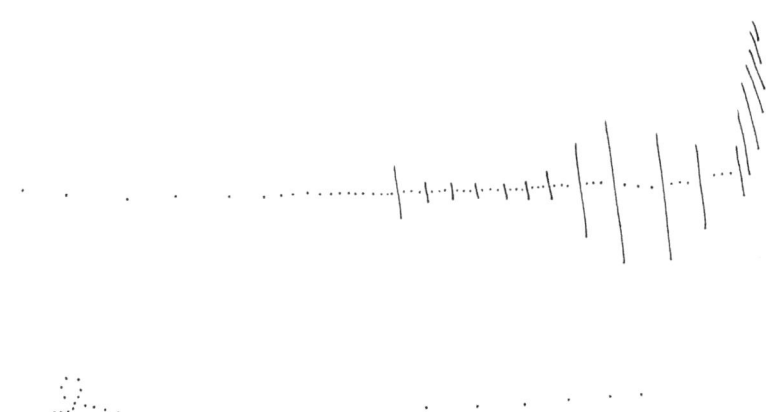

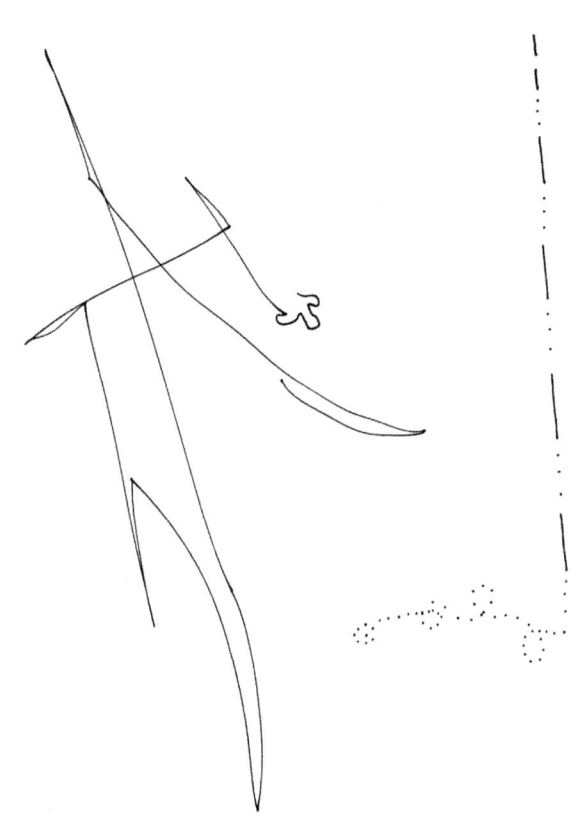

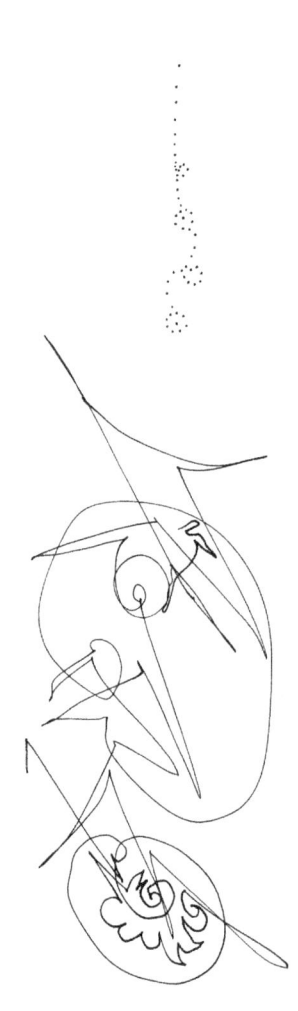

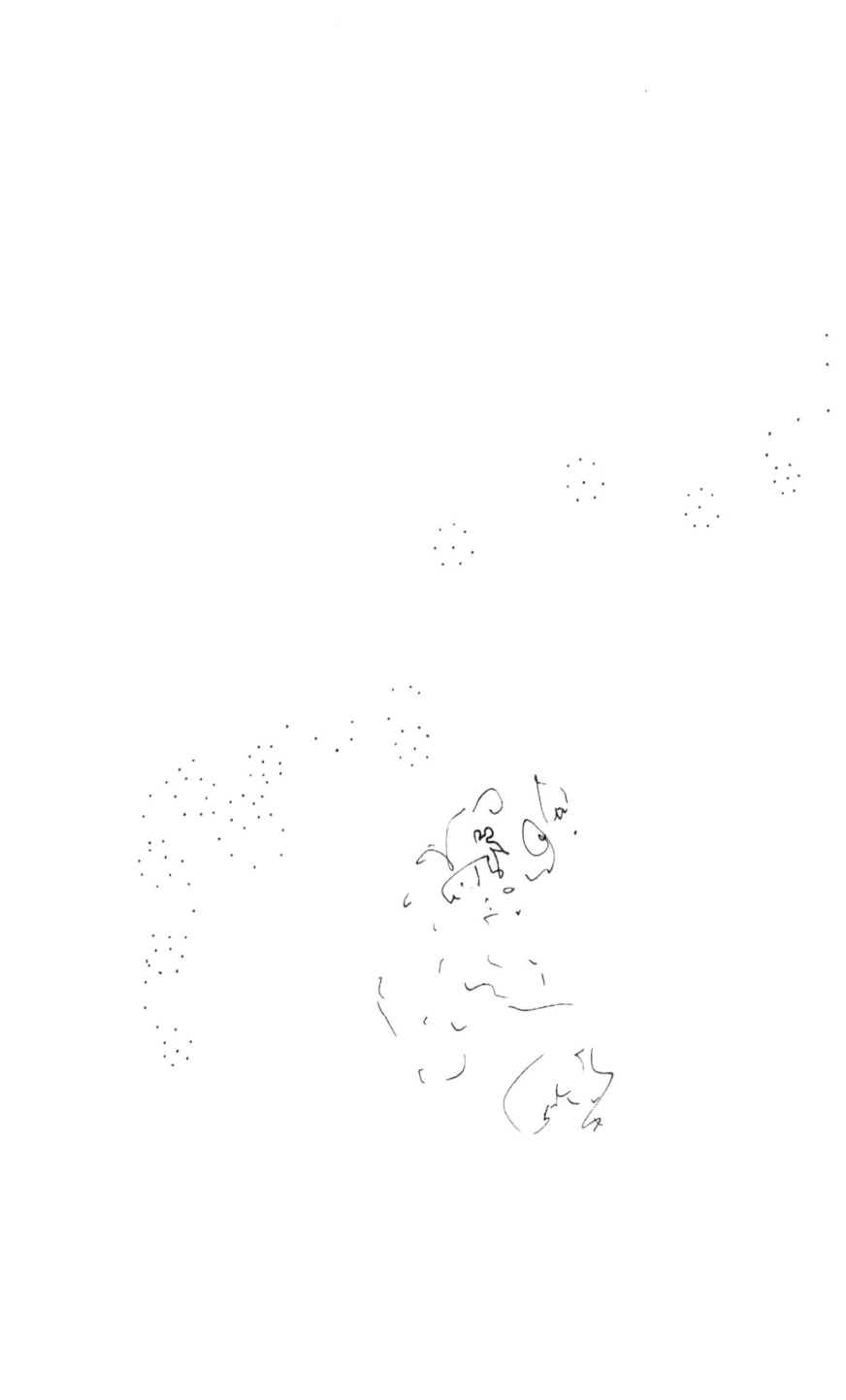

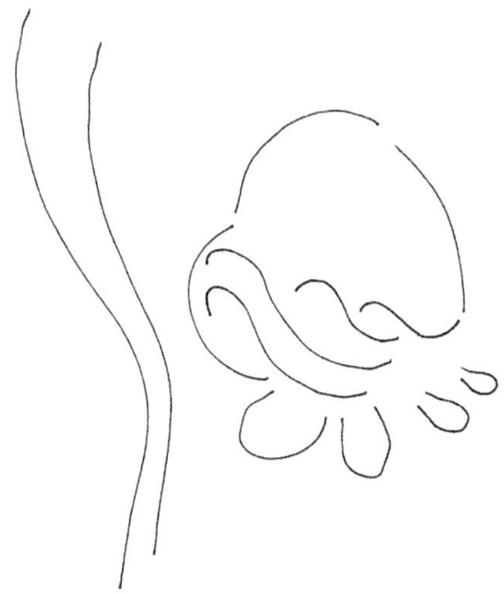

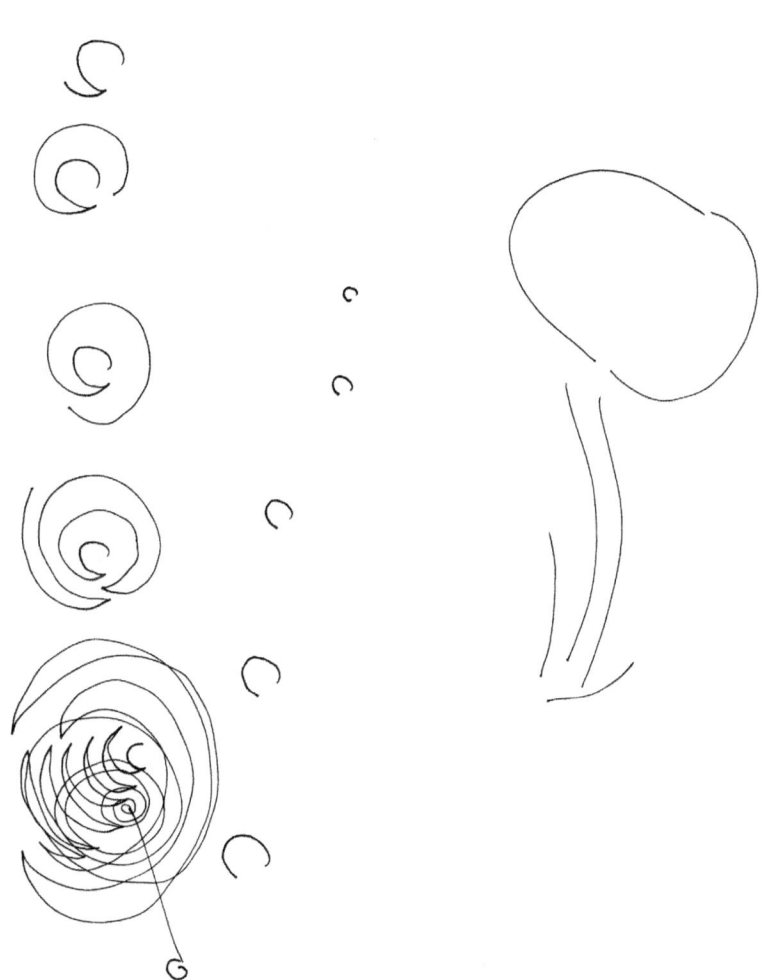

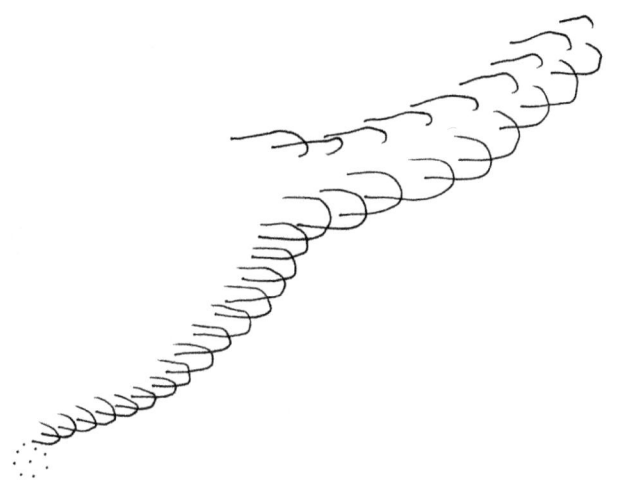

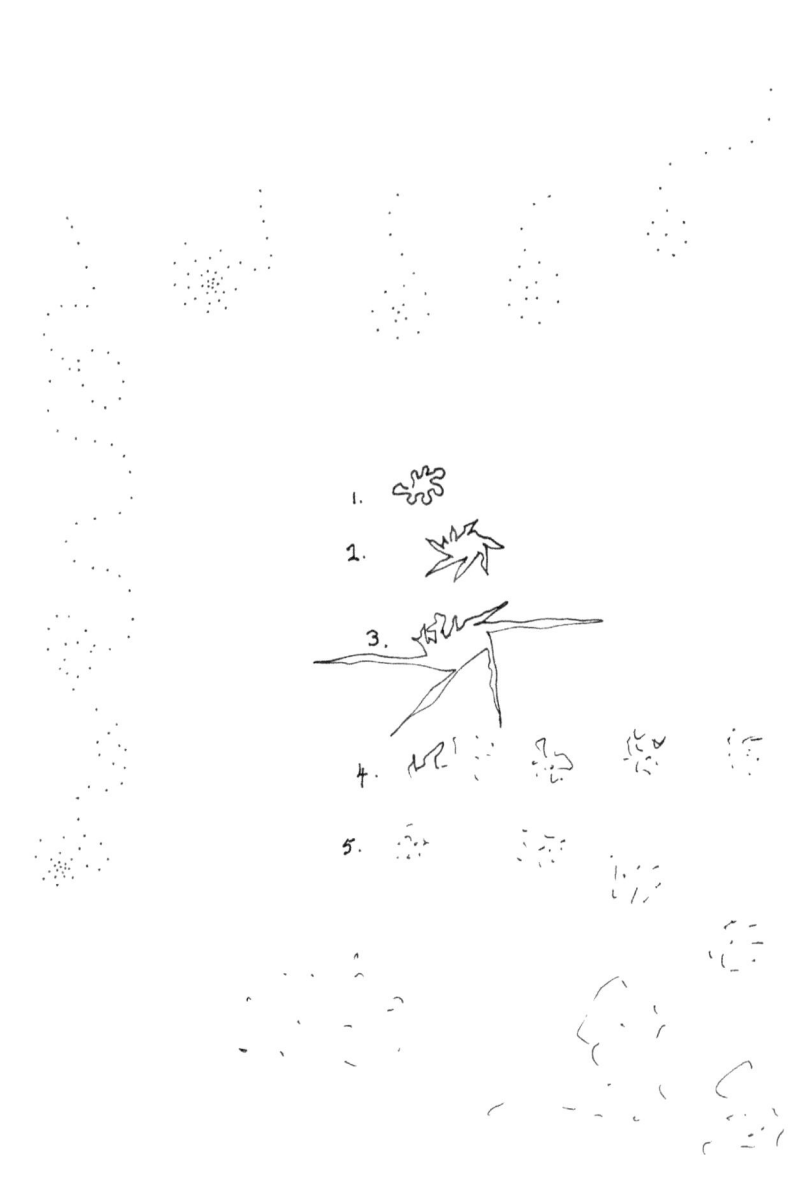

1.

1.

3.

4.

5.

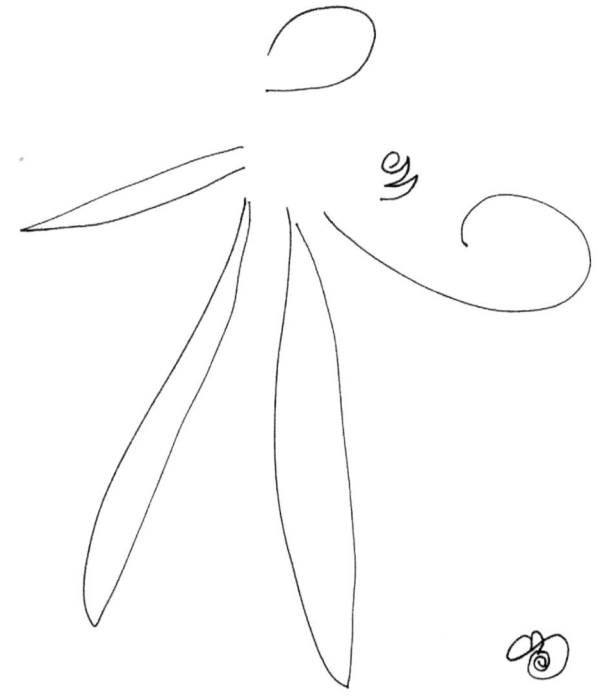

www.ingramcontent.com/pod-product-compliance
Lightning Source LLC
Chambersburg PA
CBHW080547190526
45169CB00007B/2668